ANIMALS
WE LIVE WITH

Sharon Kilzer

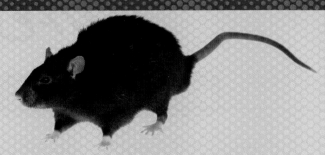

CONTENTS

Domestication: A Journey Through Time 3

DNA 13

Wild Animals Among Us 14

Displaced Animals 19

How Many Animals? 22

Glossary 24

DOMESTICATION: A JOURNEY THROUGH TIME

Go for a walk in any city in the world and you'll most likely see animals. Some of these animals are pets. Others are wild, such as pigeons and spiders. How did animals end up living with us in places like cities? We need to take a journey back through time to find out.

Imagine life about 16,000 years ago. Much of North America was covered in ice. Earth's climate was a lot colder than it is now. People lived in small groups. They hunted animals and gathered plants to eat. They drew pictures on cave walls of the animals they killed and ate.

People ate wild animals, such as horses and deer.

Then people started thinking about animals as more than just food.

From Wolf to Dog

The idea that animals could be more than something to eat began with just one animal, the wolf. Like the people of this time, wolves were living in small groups. Small groups of wolves are called packs. Like people, wolves hunted animals like horses and deer for food.

Ancient hunters and wolves competed for food.

For thousands of years, people and wolves **competed** for food, but then people started raising wolf puppies.

When the puppies grew up, most of them ran away. But some didn't. A few acted as if people were their pack. Cave drawings show that people used wolves for hunting.

With each new **generation** of puppies, people kept only the most obedient ones for **breeding**. The ones that didn't run away were tame. When these puppies grew up, they had puppies much like themselves. Gradually, over thousands of generations, people bred a new animal with this **trait**. This process of taming a wild animal and breeding it for certain traits is called domestication. People eventually gave domesticated wolves a new name—dogs.

Breeding meant that dogs ended up looking different from wolves. For example, some had shorter fur and shorter tails. But the most important thing people did was to breed dogs that behaved differently from adult wolves.

Modern dogs behave like wolf puppies. They roll on their backs like wolf puppies. They lick our hands, just like wolf puppies lick their parents' faces.

Did you know that wolves sometimes treat dogs as if dogs are young wolves?

Dogs behave the way wolf puppies do.

About 14,000 years ago, people started domesticating another animal, the caribou. Instead of hunting wild caribou, they started keeping herds of them as **livestock**.

Changing Plants

By around 12,000 years ago, people were thinking about plants in a new way, too. They thought that maybe they could change plants the way they had changed wolves and caribou.

Just as they had bred dogs from wolves, people began breeding plants such as rice and potatoes from wild plants. These early farmers started saving and planting seeds from plants that had desirable traits, such as bigger seed heads on rice. People began to grow these plants instead of gathering wild ones. This is called farming.

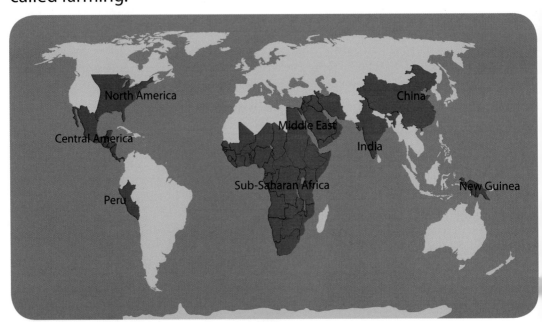

Farming developed at about the same time in different places. These are shaded red on the map.

The First Cities

With farming, people had more food. They didn't have to spend all their time hunting and gathering. With more food, more people could live together in one place. The places where more people started living together became the first towns and cities. This shift toward living in larger, **permanent** communities happened at about the same time in many different parts of the world.

One of the first cities in the world was Jericho.

Jericho may have looked something like this 10,000 years ago.

Domesticating Other Animals

As farming and cities developed, people began domesticating more and more animals. If you could go back in time and visit one of the first cities, you wouldn't find just people living in it. You would find animals, too.

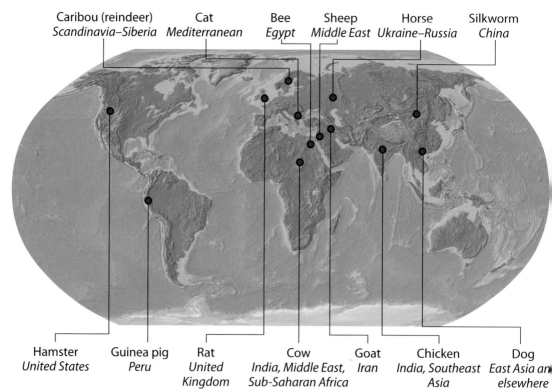

Where animals were domesticated

Caribou (reindeer) — Scandinavia–Siberia
Cat — Mediterranean
Bee — Egypt
Sheep — Middle East
Horse — Ukraine–Russia
Silkworm — China

Hamster — United States
Guinea pig — Peru
Rat — United Kingdom
Cow — India, Middle East, Sub-Saharan Africa
Goat — Iran
Chicken — India, Southeast Asia
Dog — East Asia and elsewhere

When animals w

Years ago

Dog 15,000
Caribou (reindeer) 14,000
Beginning of farming 12,000
First cities 10,000
Goat 10,000
Cat 10,000
Sheep 8,000
Cow 8,00

10

For example, in South America, people began living with guinea pigs, llamas, and alpacas.

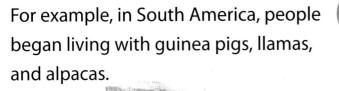
Did you know that guinea pigs aren't really a part of the pig family? They are rodents, like rats and mice. The reason why they are called guinea pigs is a mystery!

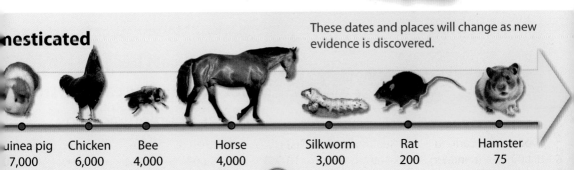

These dates and places will change as new evidence is discovered.

...nesticated							
...uinea pig	Chicken	Bee	Horse	Silkworm	Rat		Hamster
7,000	6,000	4,000	4,000	3,000	200		75

From Wild to Domestic Cats

There is evidence that one animal became domesticated without human help. When people began to grow things like wheat on farms, rats and mice moved in to eat the grain. Wild cats followed to eat the rats and mice.

! Not all wild cats are as big as lions and tigers. Some are much smaller. The desert cat is only a little bit bigger than a domestic cat. There is evidence that it is related to the domestic cat's **ancestor**.

Some domestic cats still look a bit like their wild ancestors.

Farmers realized that having cats around was a good thing. The more rats and mice the cats killed, the better! Slowly, over hundreds of years, cats became tamer. Yet even today, cats are still not as obedient as dogs.

DNA

Inside every living thing there is a **code** called DNA (which is short for deoxyribonucleic acid). This code tells your body what it should look like and how it should work. By looking at the DNA from different animals, scientists can compare their codes. If the codes are almost the same, they know that the animals are closely related. By using DNA, scientists can make a "family tree" for an animal.

The domestic dog's family tree

Grey wolf

Domestic dog

The domestic cat's family tree

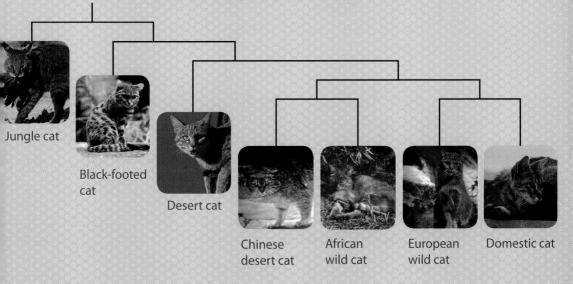

Jungle cat

Black-footed cat

Desert cat

Chinese desert cat

African wild cat

European wild cat

Domestic cat

WILD ANIMALS AMONG US

Right from the beginning, domestic animals were not the only animals living in farms, towns, and cities!

Like cats, other wild animals moved into human communities looking for food and shelter. These included hedgehogs and monkeys. Animals that use every opportunity to find food or other things they need for survival are called opportunists.

> Scientists estimate that rats scrounge 500,000 pounds of food every day in New York. That is the same as two million hamburgers a day!

Lions and Tigers

Mountain lions come into some cities looking for food. However, it is too dangerous to let mountain lions live among us. Mountain lions are taken back into the wild, and so are other dangerous animals, such as poisonous snakes.

Did you know that Korean cities once had high walls around them? The capital city of South Korea, Seoul (pronounced Soul), was called "The Fortress City" because of its 20-foot-high (6.09-meter) wall. Why did Korean cities need such high walls? To keep out tigers!

Polar Bear Jail

Towns in the Arctic have a problem. Polar bears wander into town in search of an easy meal. If you live in the Canadian Arctic and you see a polar bear in town, you need to dial the polar bear alert number. Rangers from the Ministry of Natural Resources will come and catch the bear.

How much of a problem are visiting polar bears? Well, the polar bear jail in the Canadian town of Churchill has 23 cells! It isn't really a jail, even though people call it that. The bears are rested and give water. Then they are taken back out into the wild and released. Most are young bears that have not yet learned to stay away from people. But even a young polar bear is too dangerous to have sniffing around in your yard!

Some people ask why the bears aren't fed while they are in jail. Can you guess why? It's because the rangers don't want the bears to think of Churchill as a place for a free meal. They don't want the bears to come back!

Learning to Fly in a City

Moving into a city can help save an animal from extinction. Peregrine falcons used to live only on cliffs and bluffs, but many now live in cities like Chicago, Portland, New York, and San Jose. They nest on the outside of buildings and catch pigeons to eat. With more places to live and lots of food to eat, their numbers are growing. They were recently taken off the U.S. Endangered Species List.

Thousands of people have used the Internet to watch peregrine falcon chicks hatch and learn how to fly, thanks to tiny cameras, or nestcams.

Learning to fly in a city—peregrine falcons love high buildings

DISPLACED ANIMALS

For some animals, a human community has grown up around them! Some of these animals are in the process of being **displaced**.

Snakes in Trouble

Shovel-nosed snakes live in the desert in Arizona. They "swim" through the sand using their shovel-shaped **snouts**. Shovel-nosed snakes need sand to move around in—sand is their natural habitat, or place to live. As cities like Phoenix and Tucson get larger, more and more of this snake's habitat is being lost.

Penguins Under the House

Little blue penguins are the world's smallest penguins. They nest in caves along the shore in Australia and New Zealand. When people began building houses on the shoreline, these penguins began living under them.

Nesting under a house is not as safe as nesting in a cave on a **remote** beach. Dogs kill penguins. Cars are also a problem. To get to their nests, the penguins may need to cross a road. Signs warn drivers to slow down, but little blue penguins are hard to see at night. These little birds are being displaced by coastal cities and towns.

What's it like to live with penguins? Nine-year-old Mike Dart describes what it's like to have penguins nesting under his house.

"I bet you think it would be cool to have penguins nesting under your house. It isn't! Penguins eat fish. The mother and father penguins feed the chicks by throwing up **half-digested** fish into the chicks' mouths. Some of it misses and ends up on the ground. It's gross! And the noise they make sounds like a herd of sick donkeys! No way do you want to have penguins nesting under your bedroom!"

HOW MANY ANIMALS?

How many animals now live among us? Well, it depends on what you mean by the word *animal*. Many people use *animal* to mean "all the **mammals** except us."

To a scientist, *animal* means any **multicelled** organism that lives by eating food. This includes people, fish, birds, worms, jellyfish, insects, and many, many other creatures.

In this book, we have been using the scientific meaning of the word *animal*. This means that places like cities are filled with animals—indeed, every living thing that lives in a city that isn't a plant, a fungus, or a single-celled organism is an animal. How many animals live with us? There are more than you ever imagined!

> It's not true that nests of alligators live in the sewers under New York, but did you know that someone put piranhas in the Wahiawa **reservoir**, in Hawaii, in 1992? This is *not* the way to get rid of an unwanted pet!

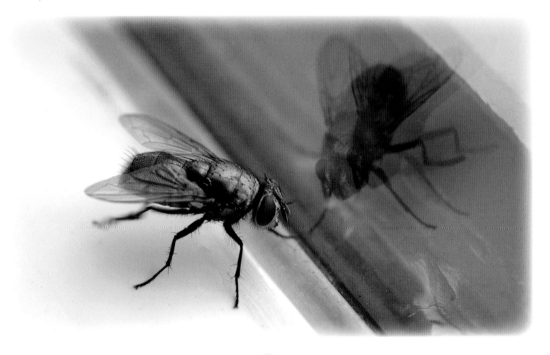

GLOSSARY

ancestor—an earlier relative
code—instructions
competed—were in a contest with each other
breeding—raising animals by selecting for traits
displaced—pushed out of a habitat
generation—a step in a line of descent
half-digested—partly processed by the body
livestock—farm animals
mammals—animals that give birth to live young and feed them with milk
multicelled—having more than one cell
permanent—existing for a very long time
remote—far away from people
reservoir—storage lake
snouts—noses, the ends of animals' faces
trait—characteristic or feature